SCIENCE IN AMERICA.

INAUGURAL ADDRESS

OF

DR. JOHN W. DRAPER,

AS PRESIDENT OF THE AMERICAN CHEMICAL SOCIETY.

DELIVERED IN CHICKERING HALL, NEW YORK, NOVEMBER 16, 1876.

Reprinted from the AMERICAN CHEMIST, *December*, 1876.

NEW YORK:
JOHN F. TROW & SON, PRINTERS.
1876.

SCIENCE IN AMERICA.

INAUGURAL ADDRESS

OF

DR. JOHN W. DRAPER,

AS PRESIDENT OF THE AMERICAN CHEMICAL SOCIETY.

DELIVERED IN CHICKERING HALL, NEW YORK,
NOVEMBER 16, 1876.

Reprinted from the AMERICAN CHEMIST, *December*, 1876.

NEW YORK:
JOHN F. TROW & SON, PRINTERS.
1876.

SCIENCE IN AMERICA.

Gentlemen, Members and Associates of the American Chemical Society:—In accordance with the plan of the American Chemical Society, I am called upon to address you this evening. I have to congratulate you on its successful establishment, and its prospect of permanent success.

Let us consider some of the reasons which would lead us to expect that success, not only for our own, but also for other kindred societies. The field of nature is ever widening before us, the harvest is becoming more abundant and tempting, the reapers are more numerous. Each year the produce that is garnered exceeds that of the preceding. In all directions there is good hope for the future. Perhaps, then, you will listen without impatience for a few minutes this evening to one of the laborers who has taken part in the toil of the generation now finishing its work, who looks back not without a sentiment of pride on what that generation has done. who points out to you the duties and rewards that are awaiting you, and welcomes you to your task. Let us look at the prospect before us.

The progress of science among us very largely depends on two elements: First, on our educational establishments. Second, on our scientific societies. To each of these I propose to direct your attention. And, first, of our colleges:

Prof. Silliman, in his address, delivered on the occasion of the Centennial of Chemistry, at the grave of Priestley, in commemoration of the discovery of oxygen, makes this remark: "The year 1845 marks the beginning of a new era in the scientific life of America, which is still in active progress, and chemistry has had its full share in this advance." He then enumerates the causes which, in his opinion, had brought about this increased activity. Among them are the Centennial celebration of the American Philosophical Society in Philadelphia in 1843; the reorganization of the United States Coast Survey in 1845; the establishment of the Smithsonian Institution at Washington in 1846; the enlargement of The American Journal of Science in the same year; the contemporaneous foundation of the Astronomical Observatory at Cincinnati; the institution of the Analytical Laboratory at Yale College in 1847, and simultaneously the Lawrence Scientific School at Harvard. To these he adds especially the establishment of the American Association for the Advancement of Science in 1848. Coinciding with him fully as to the character and power of these and other local causes which he mentions, I cannot but regard them as being themselves the issues of influences of a much more general kind.

A revolution had been taking place in Europe—a revolution not so much political as industrial and social, though it was followed by political consequences of the most important nature. Its commencements may be seen in the preceding century in the canal engineering of Brindley; in the improvements of iron manufacture; in the construction of all kinds of machinery, which reached its acme when the hand of man was deposed from its office, and through the slide-rest and planing machine engines were made by themselves. Then came the exquisite contrivances for the manufacture of textile fabrics, so that a man could do as much work in a day as he had formerly done in a year, the movement in that direction culminating in the two steam-engines, the condenser and non-condenser. The demand for cotton rose; the value of the slave, its cultivator, was enhanced; and the negro question became the paramount political question in the United States. See how scientific discoveries and inventions lead to political results! Herein, among other great events, we find the origin of the American civil war.

In Europe, the social effect of the use of steam was strikingly marked. Performing mechanical drudgery, it relieved vast numbers of the laboring class, and gave them time to think. It concentrated them in factories and mills. Those industrial hives were pervaded by literary influences, perhaps not always of a kind that we should approve of. They became the seats of agitation in politics and theology, and while this was the effect on the laboring mass, the owners or capitalists were accumulating enormous fortunes.

We may excuse the enthusiastic literature of the cotton manufacture its boasting, for men had accomplished works that were nearly godlike. Mr. Baines, writing in 1833, states that "the length of cotton yarn spun in one year was nearly five thousand million of miles—sufficient to pass round the earth's circumference more than two hundred thousand times, sufficient to reach fifty-one times from the earth to the sun. It would encircle the earth's orbit eight and a half times. The wrought fabrics of cotton exported in one year would form a girdle for the globe, passing eleven times round the equator, and more than sufficient to form a continuous sheet from the earth to the moon." And let us not forget that to give commercial value to this vast result the capital chemical discovery of bleaching by chlorine was essential. Such was the condition of things in England just previously to the epoch in question. Necessarily it was followed by great social results.

But there was something more. The locomotive absolutely revolutionized society. A man could now travel further in an hour than he had previously done in a day. Again it was clear that important political results were occurring. The effect of the railroad was to render nations more homogeneous, to destroy provincialism. It is actually true that language underwent a change. No one who had remarked the various dialects of the English counties prior to the opening of the Liverpool and Manchester Railway, and the homogeneousness of speech which is fast displacing them, could be blind to this. Simultaneously a redistribution of the population took place. It was largely withdrawn from the open country, and concentrated in the towns.

In this statement I am recalling facts so common that they are familiar to us all. We all appreciate the immense social changes that took place just before 1845. Who in those times could fail to perceive that grand consequences must follow the expenditure of thousands of millions of dollars in the building of railroads, who, when he saw the labor of a year shrinking into the compass of a day, the travel of a day into the compass of an hour, the thought of man outstripping the velocity of light, who could be so obtuse as not to discern that a new agency had taken possession of the earth, that it was agitating the nations to their very foundations, that it was ameliorating the lot of man, increasing his power, and dealing remorselessly with old ideas, the fictions and fallacies of the past!

Can we wonder, then, that those who were growing up in the midst of these marvels should not only contrast the activity by which they were surrounded with the stagnation of preceding centuries, but should demand to be made acquainted with the power that was thus opening a new world before their eyes. Very soon it became apparent that there was no provision in the existing educational establishments, the universities and colleges, for this unexpected state of things. These were, to be sure, good enough to initiate a bench of boys into the method of translating an ode of Horace or a few lines of Sophocles, but something more substantial than that was wanted now.

This was the true cause of that influence which began to be felt in America about 1840. Every reflecting person saw that a change in public education was imperative—nay, more was impending. Confronted by the vigor of modern ideas, the system that had come down from the dark ages was seen to have become obsolete.

In addition to these influences, there was another, at which we must for a moment glance. Let me, in a few words, sketch its history.

The peninsula of Italy was separated from the rule of the Greek emperors, in the eighth century, mainly in consequence of the iconoclastic dispute. Partly through the stress of circumstances, and partly as a matter of policy, the Latin language was brought into such prominence that it was supposed to contain all the useful knowledge in the world. In Western Europe, at the close of the fourteenth century, Greek was totally forgotten.

But when it became clear that Constantinople would be taken by the Turks, many learned men fled to the West, bringing with their language precious classical manuscripts. As it was feared, however, by the dominant authority that knowledge and opinions of an unsuitable kind might thus be introduced, Greek obtained a foothold with much difficulty, and it was only by the aid of Florence, Venice, and other commercial towns of Upper Italy, that after a struggle it made good its ground. The Latin had now a successful rival.

A century later brings us to the culmination of the Reformation. Its literary issue was an admiration of the language of that much enduring, that immortal race to whom the Old Testament is so largely due. As had been the case with Greek, so now Hebrew passed from a condition of neglect to one of extravagant exaltation. It was believed to have been the original language of the human race, a conviction that proved to be a great stumbling-block to the progress of learning. There were thus three classical languages, each having its own paramount claim.

In 1784 the Royal Asiatic Society was instituted in Bengal. One of its earliest and most important services was that it brought the Sanskrit language emphatically to the knowledge of Europe. The similarity of this to Latin and Greek, especially in the grammatical forms, struck every one with surprise. At first the old literary party resisted its claims, some of them even affirming that it never had been a spoken tongue, but that it had been fictitiously constructed out of Latin and Greek. The creation of comparative grammar by the great German scholar Bopp, in 1816, threw a flood of light on the subject, and the discovery in 1828 by Hodgson of the Buddhistic sacred writings in Nepaul revealed to astonished Europe a literature of grand antiquity and prodigious extent, in which is contained the religious belief of 400,000,000 of men—ten times the present population of the United States. Greek and Latin had now to descend from the imperial thrones on which they had been seated, and take their places as later and less perfect forms of this wonderful Oriental tongue.

In the higher regions of literature all over Europe these discoveries made a profound impression. It was at once seen by the great scholars of the times that the existing educational system, founded as it so largely was on the languages of the Mediterranean peninsulas, was altogether on an imperfect basis. They saw that philology was about to occupy a higher platform, and that, though it might cost a struggle with present interests, a change in public education was necessary. But though these languages have suffered an eclipse, there still remains that priceless heritage which they have transmittted to us—immortal examples in national life, in patriotism, in statesmanship, in jurisprudence, in philosophy, in poetry. Still there remain the ruins of the Parthenon, the relics of statues which have no rival elsewhere in the world—embodiments of the beautiful, before which, even at the risk of being denounced as a pagan, a man might fall down and worship. Still there remains the history of that awful empire which once bore sway around the Mediterrannean Sea, an empire to which we owe our civilization, our religious convictions, and even our modes of thought.

I add this great discovery in letters to the scientific and industrial movement I have described as bringing on the epoch of 1840.

Educational institutions are in their nature very much under the influence of the past. They are guided by men of the parting generation, and are essentially conservative. The changes they began to manifest did not originate within them, but were forced upon them from without. They clung to the mediæval as long as they could, and only accepted the modern when they were compelled.

Among American colleges which are emancipating themselves from the mediæval we may number Columbia, Cornell, Harvard, Princeton, Yale, the Universities of Pennsylvania and Virginia. Doubtless there are many others that would follow the example, if they could, but they are fettered with the gyves of sectarian or local restraint. They march along, daintily and grotesquely, in the pointed shoes of the fourteenth century.

I linger on this subject of colleges because the example of other countries, and especially of Germany, proves to us that on them our hopes for the development of science must very largely rest. The scientific glory of Germany, not inferior in brilliancy to its military glory, is the creation of its university professors.

Among them we find the great chemists and physicists, whose works we study with delight.

Our colleges must separate themselves from the mediæval, and assume thoroughly and sincerely the modern cast. Sincerely, I say, for not a few of them indulge in deception. They would have us believe that they teach physics when they have no modern apparatus; chemistry when they have no laboratory; botany without any garden, herbarium, or even drawings; geology, mineralogy, natural history, without any cabinets. So ignorant are some boards of trustees and faculties that they hold such equipments as luxuries easily dispensed with. I have known some go so far as to affirm that as much money ought to be expended in teaching a few boys Latin and Greek as in giving a demonstrative and illustrated course of science, and even to act on that principle. In institutions under this kind of influence you will always find that their whole weight is thrown towards the æsthetic. Whatever college honors there may be, whatever emoluments, pass in that direction; and though through fear of public opinion science cannot be ignored, it is simply tolerated, not cultivated.

From our colleges we may, in the second place, turn to our scientific societies.

I have referred to the period at which the Greek language became cultivated in Western Europe. The first societies were those established in Florence by its admirers. In the Medicean gardens the lovers of Plato assembled to restore, under an Italian sky, the philosophy that had been extinguished in Athens, and to commemorate by a symposium the birthday of that illustrious man. There is a pleasure in associating with those whose thoughts are congenial to our own, in breathing an atmosphere in which the intellectual makes itself felt.

Very soon the example was imitated. Persons who

had a love for science followed the example of those who had a love for letters. The Academia Secretorum Naturæ was instituted at Naples, in 1560, by Baptista Porta, the inventor of the camera, which photographers now so much use; the Lyncean Academy for the Promotion of Natural Philosophy, in 1603; the Royal Society of London, 1645; the Royal Academy of Science in Paris, 1666; the Berlin Academy of Arts and Sciences, in 1700. Leibnitz, the rival of Newton, was its first President.

When the Royal Society of London was founded, it encountered a bitter opposition. Had it not been for the "Merry Monarch," Charles II., it must have succumbed beneath the fierce maledictions launched against it.

As in Italy, when the opportunity was offered, men of the same inclination of thinking sought each other, so here, to the surprise of the most enthusiastic chemists, when such an association was proposed persons seeking membership came crowding in. The society I have the honor of addressing this evening was the result. Already it has completely organized itself; already it has published the first number of its "proceedings," a publication which I am sure will procure for it approval and respect.

In these organizations of scientific effort an opportunity of assisting is given to those who, not having dedicated themselves to philosophical pursuits, have yet achieved success in other walks of life, and who, recognizing that the progress of civilization very largely depends on the increase of knowledge, may desire to aid in promoting that great result by the application of their means. See what immense benefits have arisen from the money grants that foreign governments have placed at the disposal of their scientific bodies; see what a stimulus there has been in the award of medals of honor, and, if you desire to witness the effect of a well-judged benefaction, look at the Smithsonian Institution. I would not say one word in disparagement of gifts to colleges and universities, for it is indeed a noble purpose; but endowments for the promotion of a knowledge of nature conferred on scientific societies for the good of all men, no matter what their country or color, no matter what their religious profession or political condition, are still nobler. The one is a local and transitory benefaction, the other an enduring and universal benevolence.

In our own special science, chemistry, all that has been done has only served to extend the boundary of what remains. The thousands of analyses that have been made have brought us into a wilderness of results. We have not been able to rise to a point of view sufficiently high to discover what is the true place of those results in nature. We try to represent on the pages of our books and on our blackboards formulas of the constitution of things, conscious all the time that these are at the best only convenient fictions, which must necessarily change as we gain a more perfect insight into that grandest of all problems, the distribution of Force in Space, and the variations to which it is liable. The geometry of chemistry is that of three dimensions, not of two. We have to consider the relation of points not situated on one plane, and hence it is necessary to employ three axes of reference; nay, even more, we cannot avoid the conception of the mathematical method of quaternions. Our inadequate information respecting the real grouping of atoms is followed, as a necessary consequence, by imperfection in our methods of nomenclature—the confusion in this respect becoming, as we all too well know, every day worse and worse.

And now, while we have accomplished only a most imperfect examination of objects that we find on the earth, see how, on a sudden, through the vista that has been opened by the spectroscope, what a prospect lies beyond us in the heavens! I often look at the bright yellow ray emitted from the chromosphere of the sun, by that unknown element, Helium, as the astronomers have ventured to call it. It seems trembling with excitement to tell its story, and how many unseen companions it has. And if this be the case with the sun, what shall we say of the magnificent hosts of the stars? May not every one of them have special elements of its own? Is not each a chemical laboratory in itself? Look at the clusters in the sword-handle of Perseus; in Cassiopeia, a universe of stars on a ground of star-dust; in Hercules, of which, as astronomers say, no one can look at for the first time through a great telescope without a shout of wonder—the most superb spectacle that the eye of man can witness! Look at the double stars, of which so many are now known, emitting their contrasting rays, garnet, or ruby, or emerald, or sapphire? Each is in accordance with its own special physical conditions, though all are under the same universal ordinance.

Now here a fact of surpassing importance presses itself on our attention. The movements taking place in those distant bodies are taking place under the same laws that prevail here on earth, and in our solar system. The law of gravitation, as developed by Newton, bears sway in all these distant worlds. In them bodies attract each other with forces directly as their masses and inversely as the squares of their distances. There the laws of the emission, absorption, and transmission of light are the same as they are with us. There ignited hydrogen gives forth its three rays, the same rays that it gives forth to us. In the uttermost parts of the universe the law of definite combination, the numerical law, and the multiple law, stand good. Sodium absorbs its two waves of definite refrangibility, and iron gives in the spectra its more than a hundred lines—more than a hundred silent but convincing witnesses of the uniformity of the constitution of the universe. There the number of vibrations that constitute a ray of definite refrangibility is the same we have found it to be here. In the enormous heat of those central suns, the dissociation of molecules may be of a higher order than we can reach artificially, but the law under which it takes place is a continuation of the law here. There, though the weight of a given mass of matter is different from what it is with us, it is nevertheless determined by the law that determines it here—the law of gravitation. There energy is indestructible, and is measured as it is measured among us, by work. Then is there any boundary that we can assign to natural law—is it not omnipresent, universal?

Perhaps there is no exaggeration in the assertion—for there seems abundant proof of its truth—that the light by which we see some of those distant orbs has crossed through such a prodigious space that millions of years have transpired during the journey. Then the phenomena it brings to us are those that were engendered in the beginning of the vast time so passed. Whatever there is that is in harmony with facts now happening here, is to us an unimpeachable evidence that the laws which were governing in those old ages have undergone no depreciation, but are active as ever until now. Then shall I exaggerate if I say that those laws are eternal in duration?

Infinite in influence, eternal in duration! what a magnificent spectacle! In the resistless energy of the motions of the universe, is there not Omnipotence? The Omnipotent, the Infinite, the Eternal—to what do these attributes belong?

Shall a man who stands forth to vindicate the majesty of such laws be blamable in your sight? Rather shall you not with him be overwhelmed with a conception so stupendous? And yet let us not forget that these eternal laws of nature are only the passing thoughts of God.

But grand as this is, there is something still grander. There is another temple into which we have to pass—not that of the visible, but that of the invisible. We must persist in the invasion we have made, in the revolution we have brought about in physiology. We have to determine the laws which preside in the nervous system of man, and discover the nature of the principle that animates it. Is there not something profoundly impressive in this, that the human mind can look from without upon itself, as one looks at his phantom image in a mirror and discern its own lineaments and admire its own movements. My own thoughts have of late years been forcibly drawn to this, from a recognition that the interpretation by the mind of impressions from without takes place under mathematical laws; as, for instance, that when external ethereal vibrations create in the mind a certain idea, that same idea will arise when the vibrations are doubled, or tripled, or quadrupled in frequency; but other ideas will be engendered by vibrations of an intermediate rate. Yet what these ideas will be may be predicted. It is true that this is only an optical case, but it extends the view that has been offered to us by a study of the structure of the ear. In the labyrinthine compartment of that organ the ultimate fibres of the auditory nerve are laid on the winding plane of the spiral lamina, in ever-decreasing lengths, each capable of trembling to the sound which is in unison with it—a mechanical action truly, answering to the sympathetic vibration with which the strings of a piano will respond to the corresponding notes of a flute—and these are translated by the mind into all the utterances of articulate speech, all the harmonies of music—speech that engenders new ideas within us; strains which, though they may die away in the air, live forever in the memory. The exquisite delight we experience in listening to the works of our great composers arises thus in mechanical movements, which are the issue of mathematical combinations. The unseen world is under the influence of number!

But what is number except there be one who numbers? When Pompey, in his Syrian war, broke into the holy of holies at Jerusalem, he expressed, as Tacitus tells us, his astonishment that there was no image of a Divinity within—the shrine was silent and empty. And so, though after death we may anatomize and explore the inmost recesses of the brain, the vailed Genius that once presided there has eluded us, and has not left so much as a phantom trace, a shadow of himself.

The experiments of Galvani and Volta have not yet reached their conclusion; those of Faraday and Du Bois Raymond have only yielded a preliminary suggestion as to the nervous force. Excepting the great sympathetic nerve, the nervous fibres themselves are, as is well known, of two classes—those that gather the impressions of external things and convey them to the nerve centres, and those that transmit the dictates of the will from within outwardly. The capabilities of one of the former—the apparatus for sight—have been greatly improved by various optical contrivances, such as microscopes and telescopes, an earnest of what may hereafter be done as respects the four other special organs of sense; and as concerns the second class, the result of mental operations, the resolves of the will, may be transmitted with greater velocity than even in the living system itself, and that across vast terrestrial distances, or even beneath the sea. Telegraphic wires are, strictly speaking, continuations of the centrifugal nerves, and we are not without reason for believing that it is the same influence which is active in both cases.

In a scientific point of view, such improvements in the capabilities of the organs for receiving external impressions, such extensions in the distances to which the results of intellectual acts and the dictates of the will may be conveyed, constitute a true development, an evolution, none the less real though it may be of an artificial kind. If we reflect carefully on these things, bearing in mind what is now known of the course of development in the animal series, we shall not fail to remark what a singular interest gathers round these artificial developments—artificial they can scarcely be called, since they themselves have arisen interiorly. They are the result of intellectual acts Man has been developing himself. He, so far as the earth is concerned, is becoming omnipresent. The electrical nerves of society are spread in a plexus all over Europe and America; their commissural strands run under the Atlantic and the Pacific.

In many of the addresses that have been made during the past summer, on the Centennial occasion, the shortcomings of the United States in extending the boundaries of scientific knowledge, especially in the physical and chemical departments, have been set forth. "We must acknowledge with shame our inferiority to other people," says one. "We have done nothing," says another. Well, if all this be true, we ought, perhaps, to look to the condition of our colleges for an explanation. But we must not forget that many of these humiliating accusations are made by persons who are not of authority in the matter—who, because they are ignorant of what has been done, think that nothing has been done. They mistake what is merely a blank in their own information for a blank in reality. In their alacrity to depreciate the merit of their own country, a most unpatriotic alacrity, they would have us confess that for the last century we have been living on the reputation of Franklin and his thunder-rod.

Perhaps, then, we may without vanity recall some facts that may relieve us in a measure from the weight of this heavy accusation. We have sent out expeditions of exploration both to the Arctic and Antarctic seas. We have submitted our own coast to a hydrographic and geodesic survey not excelled in exactness and extent by any similar works elsewhere. In the accomplishment of this we have been compelled to solve many physical problems of the greatest delicacy and highest importance, and we have done it successfully. The measuring rods with which the three great base lines of Maine, Long Island, and Georgia were determined, and their beautiful mechanical appliances, have exacted the publicly expressed admiration of some of the greatest European philosophers, and the conduct of that survey their unstinted applause. We have instituted geological surveys of many of our States and much of our Territories, and have been rewarded not merely by manifold local benefits, but also by the higher honor of extending very greatly the boundaries of that noble science. At an enormous annual cost we have maintained a meteorological signal system which I think is not equalled, and certainly is not surpassed, in the world. Should it be said that selfish interests have been mixed up with some of these undertakings, we may demand whether there was any selfishness in the survey of the Dead Sea? Was there any selfishness in the mission that a citizen of New York sent to equatorial Africa for the finding and relief of Livingstone? any in the astronomical expedition to South America? any in that to the valley of the Amazon? Was there any in the sending out of parties for the observation of the total eclipses of the sun? It was by American astronomers that the true character of his corona was first determined. Was there any in the seven expeditions that were dispatched for observing the transit of Venus? Was it not here that the bi-partition of Biela's comet was first detected; here that the eighth satellite of Saturn was discovered; here that the dusky ring of that planet, which had escaped the penetrating eye of Herschel and all the great European astronomers, was first seen? Was it not by an American telescope that the companion of Sirius, the brightest star in the heavens, was revealed, and the mathematical prediction of the cause of his perturbations verified? Was it not by a Yale College professor that the showers of shooting stars were first scientifically discussed, on the occasion of the grand American display of that meteoric phenomenon in 1833? Did we not join in the investigations respecting terrestrial magnetism, instituted by European governments at the suggestion of Humboldt, and contribute our quota to the results obtained? Did not the Congress of the United States vote a money grant to carry into effect the invention of the electric telegraph? Does not the published flora of the United States show that something has been done in botany? Have not very important investigations been made here on the induction of magnetism in iron, the effect of magnetic currents on one another, the translation of quantity into intensity, and the converse? Was it not here that the radiations of incandescence were first investigated, the connection of increasing temperature with increasing refrangibility shown; the distribution of light, heat, and chemical activity in the solar spectrum ascertained, and some of the fundamental facts in spectrum analysis developed long before general attention was given to that subject in Europe? Here the first photograph of the moon was taken; here the first of the diffraction spectrum was produced; here the first portraits of the human face were made—an experiment that has given rise to an important industrial art?

Of chemistry, it may truly be affirmed that nowhere are its most advanced ideas, its new conceptions, better understood or more eagerly received. But how useless would it be for me to attempt a description in these few moments of what Prof. Silliman, in the work to which I have already referred, found that he could not include on more than

100 closely printed pages, though he proposed merely to give the names of American chemists and the titles of their works. It would be equally useless and indeed an invidious task to offer a selection; but this may be said, that among the more prominent memoirs there are many not inferior to the foremost that the chemical literature of Europe can present. How unsatisfactory, then, is this brief statement I have made of what might be justly claimed for American science! Had it been ten times as long, and far more forcibly offered, it would still have fallen short of completeness. I still should have been open to the accusation of not having done justice to the subject.

Have those who gloat over the shortcomings of American science ever examined the Coast Survey reports, those of the Naval Observatory, the Smithsonian contributions, those of the American Association for the Advancement of Science, the proceedings of the American Academy of Arts and Science, those of the American Philosophical Society, the Lyceum of Natural History, and our leading scientific periodicals? Have they ever looked at the numerous reports published by the authority of Congress on geographical, geological, engineering, and other subjects—reports often in imposing quartos magnificently illustrated.

Not without interest may we explore the origin of the depreciation of which we thus complain. In other countries it is commonly the case that each claims for itself all that it can, and often more than is its due Each labors to bring its conspicuous men and its public acts into the most favorable point of view; each goes upon the maxim that a man is usually valued at the value he puts upon himself. But how is it with us? Can an impartial person read without pain the characters which we so often attribute to our most illustrious citizens in political, and, what is worse, in social life? Can we complain if strangers accept us at our own depreciation, whether of men or things?

We need not go far to detect the origin of all this —it is in our political condition. Here wealth, power, preferment—preferment even to the highest position in the nation—are seemingly within the reach of all, and in the internecine struggle that takes place every man is occupied in pushing some other man into the background. I fear that in political life there is no remedy for this, such is the violence of the competition, so great are the prizes at stake. But in the less turbulent domain of science and letters we may hope for better things. And those who make it their practice to decry the contributions of their own country to the stock of knowledge may perhaps stand rebuked by the expressions that sometimes fall from her generous rivals. How can they read, without blushing at their own conduct, such declarations as that recently uttered by the great organ of English opinion, the foremost of English journals! *The Times*, which no one will accuse of partiality in this instance, says: "In the natural distribution of subjects, the history of enterprise, discovery, and conquest, and the growth of republics fell to America, and she has dealt nobly with them. In the wider and multifarious provinces of art and science she runs neck and neck with the mother country, and is never left behind?"

There are among us some persons who depreciate science merely through illiterate arrogance; there are some who, incited by superficiality, dislike it; there are some who regard it with an evil eye, because they think it is undermining the placid tranquillity they find in lifelong cherished opinions. There are some who hate it because they fear it, and many because they find that it is in conflict with their interests.

But let us, who are the servants of science, who have dedicated ourselves to her, take courage. Day by day the number of those who hold her in disfavor is diminishing. We can disregard their misrepresentations and maledictions. Mankind has made the great discovery that she is the long-hoped-for civilizing agent of the world. Let us continue our labor unobtrusively, conscious of the integrity of our motives, conscious of the portentous change which is taking place in the thought of the world, conscious of the irresistible power which is behind us! Let us not return railing for railing, but above all, let us deliver unflinchingly to others the truths that Nature has delivered to us!

The book of Nature! shall not we chemists, and all our brother-students, whether they be naturalists, astronomers, mathematicians, geologists, shall we not all humbly and earnestly read it? Nature, the mother of us all, has inscribed her unfading, her eternal record on the canopy of the skies, she has put it all around us on the platform of the earth! No man can tamper with it, no man can interpolate or falsify it for his own ends. She does not command us what to do, nor order us what to think. She only invites us to look around. For those who reject her she has in reserve no revenges, no social ostracism, no index expurgatorius, no auto da fé! To those who in purity of spirit worship in her heaven-pavilioned temple, she offers her guidance to that cloudy shrine in which Truth sits enthroned, "dark with the excess of light?" Thither are repairing, not driven by tyranny, but of their own accord, increasing crowds from all countries of the earth, conscious that whatever their dissensions of opinion may heretofore have been, in her presence they will find intellectual concord and unity.

www.ingramcontent.com/pod-product-compliance
Lightning Source LLC
LaVergne TN
LVHW011141110826
845150LV00008B/2459
* 9 7 8 1 4 1 8 1 9 3 3 1 7 *